FLORA OF TROPICAL EAST AFRICA

TACCACEAE

Susan Carter

Tuberous, perennial herbs. Leaves erect, large, basal, petiolate, the base of the petiole channelled or sheathing ; leaf-blade entire, palmate, digitate or pinnatisect. Flowering stems 1–3, basal, erect, simple, leafless. Inflorescence terminal, umbellate with numerous, pedicellate flowers usually surrounded by a whorl of filiform bracts, and 4–12 large, foliaceous bracts on the outside. Flowers bisexual. Perianth-segments 6, petaloid, connate at the base, persistent. Stamens 6, adnate to the perianth-segments ; filaments hooded and lobed ; anthers 2-celled, dehiscing longitudinally, introrse, situated on the inside of the hood, and projecting outwards between the lobes. Ovary inferior, unilocular with 3 parietal placentas ; ovules many ; style short ; stigma 3-lobed and umbrella-shaped with the stigmatic surface beneath. Fruit capsular or baccate. Seeds many, with copious endosperm.

A family of 2 genera and perhaps 20 species, mostly from tropical Asia. The other genus, *Schizocapsa* Hance, with capsular fruit, is monotypic and recorded from eastern Asia only.

TACCA

J.R. & G. Forst., Char. Gen. : 69 (1776)

Glabrous or sparsely pilose herbs. Tuber subglobose. Bracts green or coloured, the 2 outer sessile, the inner sessile or scarcely petiolate ; filiform bracts usually present. Perianth-segments subequal, reflexed, spreading, or erect. Stigma-lobes emarginate. Fruit baccate. Seeds reniform or ovate ; embryo minute, buried in the endosperm.

T. leontopetaloïdes (*L.*) *O. Ktze.*, Rev. Gen. 2 : 704 (1891) ; Troupin, Fl. Parc Nat. Garamba 209 (1956). Type : India ; Tab. & descr. of *Leontopetaloïdes* by J. Amman in Comm. Acad. Imp. Sci. Petrop. 8 : 211, t. 113 (1736) (holo. !)

Stout herb, up to 1·5 m. Tuber up to 10 cm. across ; roots fibrous. Leaves 1–3 ; sheathing cataphylls 2 or 3, up to 5 cm. long, enclosing the bases of the petioles and flowering stems ; petiole ridged longitudinally, up to 1 m. long, with a sheathing base ; leaf-blade trisected at the apex of the petiole, each segment 1–3-pinnatisect ; lobes ovate-acuminate, up to 12 × 7 cm., and with smaller orbicular lobes between them up to 3 × 2·5 cm. Flowering stems usually 1, ridged longitudinally, up to 1·5 m. high ; outer bracts 2, spreading, ovate-lanceolate, usually with 2–3-fid tips, (2–)4(–7) × (1–)2·5(–4) cm. ; inner bracts 3–4, erect, ovate, almost sessile, slightly shorter than the outer ones ; both types green tinged with purple ; filiform bracts many, purplish with whitish tips, stiff and pendulous, up to 20 cm. long ; flowers 20–40, but only a few of these produce fruit ; pedicels about 2·5 cm. long, elongating to about 4 cm. in fruit. Perianth-segments erect, green tinged with purple, thickened ; outer segments

1

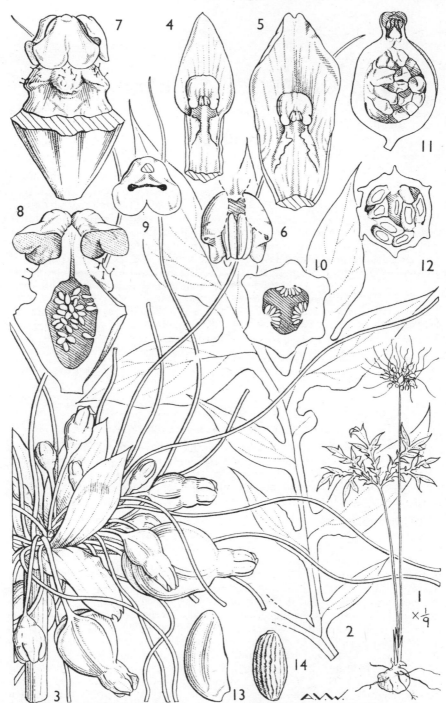

FIG. 1. *TACCA LEONTOPETALOÏDES*—**1**, plant, × 1/9 ; **2**, leaf, × 2/3 ; **3**, inflorescence, × 1 ; **4**, outer per. seg., × 4 ; **5**, inner per. seg., × 4 ; **6**, stamen from inside hood, × 8 ; **7**, ovary and style, × 6 ; **8**, longitudinal section of same, × 6 ; **9**, stigma-lobe from below, × 6 ; **10**, transverse section of ovary, × 6 ; **11**, longitudinal section of mature fruit, × 1 ; **12**, transverse section of mature fruit, × 1 ; **13**, seed surrounded by the aril, × 3 ; **14**, seed, × 3. 1, from *Milne-Redhead & Taylor 8263*, 8263A, and a drawing by *H. Faulkner* ; 2, from *Milne-Redhead & Taylor 8263* ; 3–10, from *Milne-Redhead & Taylor 8263A* ; 11–14, from *Milne-Redhead & Taylor 8263B*.

oblong, 6 × 3·5 mm. ; inner segments oblong, emarginate, 8 × 6 mm. ;
perianth-tube 4 mm. long. Filaments 2·5 mm. wide, adnate to the perianth-
tube but with free margins ; hood 2 × 2 mm. ; anthers 2 mm. long.
Style 2 mm. long, set upon a basal swelling covered with small stalked
glands ; stigma-lobes heart-shaped, 3 × 2·5 mm. ; stigmatic surface a
slit, 1·5 mm. long ; ovary obovate, 4 × 3 mm. Fruit subglobose, 6-ridged,
up to 3 × 2·5 cm. Seeds ovate, longitudinally ridged, 5 × 3 mm., red-
brown, but surrounded by a thin fleshy aril which is not evident when the
seeds are dry. Fig. 1.

UGANDA. West Nile District : Terego, Apr. 1938, *Hazel* 497 ! ; Bunyoro District :
Masindi, 12 May 1941, *A. S. Thomas* 3888 ! ; Teso District : Serere, Apr.-May 1932,
Chandler 571 !
KENYA. Mombasa, English Point, 26 May 1935, *Napier* 6294 ! ; Kilifi District : Mida,
May 1929, *R. M. Graham* 2120 !
TANGANYIKA. Lushoto District : Korogwe, Magunga, 26 May 1953, *Faulkner* 1192 ! ;
Ufipa District : near Lake Rukwa, 7 Jan. 1938, *Michelmore* 1454 ! ; Songea District :
7 km. W. of Songea, 18 Jan. 1956 (fl.), *Milne-Redhead & Taylor* 8263 ! & 6 Feb. 1956
(fr.), 8263A !
ZANZIBAR. Zanzibar I., Apr. 1874, *Hildebrandt* 1285 ! ; Donge, 20 June 1950, *Oxtoby*
18 ! ; Pemba I., Chake-Chake, *Vaughan* 309 !
DISTR. U1-4 ; K7 ; T3, 4, 6-8 ; Z ; P ; tropical and subtropical ; from India and
China through Malaya to the Pacific islands and north Australia, Madagascar and
the Mascarene Is., and in Africa from Sierra Leone to Ethiopia and southwards to
Southern Rhodesia
HAB. Grassland, bushland, or woodland usually on sandy soils ; 0–1100 m.
SYN. *Leontice leontopetaloïdes* L., Sp. Pl. : 313 (1753)
   Tacca pinnatifida J. R. & G. Forst., Char. Gen. : 70, t. 35 (1776) ; F.T.A. 7 :
      413 (1898). Type : Tahiti, *Forster* (K, holo. !)
   T. involucrata Schumach. & Thonn., Beskr. Guin. Pl. : 177 (1872) ; F.W.T.A.
      2 : 397, fig. 320 (1934). Type : Ghana, *Thonning* (C., holo.†)
   T. quanzensis Welw., Apont. : 591 (1858). Type : Angola, Pungo Andongo,
      Welwitsch 6475 (BM, holo. !)
   T. pinnatifida J. R. & G. Forst. subsp. *madagascariensis* Limpr. f., Beitr.
      Kenntn. Tacc. Diss. : 53 (1902). Type : Ethiopia, *Schimper* 1946 (K,
      syn. !). Syntypes also from Senegal and Portuguese East Africa
   T. pinnatifida J. R. & G. Forst. subsp. *involucrata* (Schumach. & Thonn.)
      Limpr. f., Beitr. Kenntn. Tacc. Diss. : 55 (1902)
   T. pinnatifida J. R. & G. Forst. f. var. *acutifolia* Limpr. f., Beitr. Kenntn. Tacc. Diss. : 55 (1902).
      Syntypes from Ethiopia, Tanganyika and Nyasaland
   T. umbrarum Jum. & Perr. in Ann. Mus. Colon. Mars., sér. 2, 8 : 386, t. 1, 2
      (1910). Type : Madagascar, *Jumelle* (location not known)
   T. madagascariensis (Limpr. f.) Limpr. f. in E.P. IV. 42 : 29 (1928)
   T. involucrata Schumach. & Thonn. var. *acutifolia* (Limpr. f.) Limpr. f. in E.P. IV.
      42 : 29 (1928)

VARIATION. Various leaf-forms of this plant have been described as distinct species.
The size and shape of the lobes, however, appear to depend on environmental con-
ditions, the plants growing in dry, open situations having many small narrow lobes,
while those with a few large lobes are found in relatively damp, shady places.

INDEX TO TACCACEAE